Bibliographic information published by the German National Library:

The German National Library lists this publication in the National Bibliography; detailed bibliographic data are available on the Internet at http://dnb.dnb.de .

Imprint:

Print and binding: Books on Demand GmbH, Norderstedt Germany
ISBN: 9783668380059

This book at GRIN:

http://www.grin.com/en/e-book/351143/novel-and-innovative-techniques-in-meli-poniculture-in-kerala-the-current

Prem Jose Vazhacharickal, Sajan Jose K

Novel and innovative techniques in Meliponiculture in Kerala. The current status of knowledge

GRIN Publishing

Novel and innovative techniques in Meliponiculture in Kerala: current status of knowledge

ACKNOWLEDGEMENT

Firstly we thank **God Almighty** whose blessing were always with us and helped us to complete this research work successfully.

The first author is extremely grateful to **Dr. Sajeshkumar N.K** (Head of the Department, Biotechnology) for the valuable suggestions, support and encouragements.

The first author would wish to thank beloved Manager **Rev. Fr. Dr. George Njarakunnel,** respected Principal **Dr. V.J. Joseph,** Bursar **Shaji Augustine,** Vice Principal **Fr. Joseph Allencheril**, and the Management for providing all the necessary facilities in carrying out the study.

We lovingly and gratefully indebted to our teachers, parents, siblings and friends who were there always for helping us in this project.

Table of contents

Table of figures

Table of tables

List of abbreviations

LED	: Light emitting diode
N	: Nectar
NP	: Nectar and pollen
P	: Pollen
PVC	: Polyvinyl chloride

Abstract

Stingless bees are highly social insects which populated the tropical earth 65 million years ago longer than honey bees. They are limited to tropics and subtropics lacking venom apparatus and cannot sting. A little is reported so far about the domestication of the *Trigona iridipennis* Smith as well as the aspects of Meliponiculture in Kerala. Based on these back ground, the objectives of this study were to 1) to characterize the various Meliponiculture techniques and management practices 2) the new techniques evolved for Meliponiculture practices and management as well as value addition of stingless bee products. Hundred and twenty farmers across Kerala engaged in Meliponiculture were interviewed and their Meliponiculture practices were observed and recorded. A questionnaire was prepared and distributed among the farmers to learn their depth of knowledge in stingless bee keeping techniques and to study the current status of Meliponiculture. In earthen bowl and coconut shell methods it is very advantageous to collect honey without causing any disturbance to the brood chamber that is at the bottom. In PVC pipe and coconut shell the colony often desert in summer due to high temperature. Even though the bamboo nodes are good, they are found to degenerate after few years through the attack of wood borers. Stingless bee rearing is an important activity as it helps pollination and provides honey for various purposes. The bottle neck for Meliponiculture includes availability of colonies to start with, better techniques for colony capture, colony multiplication, honey extraction, processing and marketing. This study highlights the various drawbacks and flaws of traditional Meliponiculture in Kerala, which could be rectified using improved scientific management practices and tools. If correctly informed disseminated these techniques could be utilized by tribal, woman, old aged people and even children to fetch an extra income and also to serve the high demand of stingless bee honey and help to maintain the rich biodiversity of Kerala

Keywords: *Trigona iridipennis* Smith; Anthropogenic habitats; Meliponiculture; Eduction techniques

1. Introduction

The art and science of keeping stingless bees for honey, pollen, resin, ecological services is called Meliponicultre. This has been practiced even during Mayan culture and in different parts of Latin America (Kent, 1984; Sommeijer et al., 2003; Ayala et al., 2013). In Costa Rica, Meliponiculture is practiced in two ways: 1) maintain colonies in tree trunks, from which honey is extracted through a lateral opening and 2) keeping the colonies in small wooden boxes, bamboo poles or in hollow gourds (Van Veen et al., 2000). In Brazil, for most people it is a fast growing secondary economic activity (Cortopassi-Laurino et al., 2006). *Melipona bocandei*, the largest stingless bee in Angola produces 10-15 kg honey per year (Armor, 2005). The stingless bee keepers usually keep their colonies in log hives, generally hanging from the eves of their houses. Special places for meliponiary are built away from the houses (Cortopassi-Laurino et al., 2006). Stingless bees are effective pollinators, and keeping them for pollination services is now beginning to take root in southern Asia (India) and South-East Asia (Malaysia and Philippines) (Heard, 1994; Heard, 1999; Cortopassi-Laurino et al., 2006). In Kerala, stingless bee rearing is a traditional activity in rural households without going into the science or economics. Compared to Apiculture this is not vigorously promoted. But in the changing eco-economic scenario, Meliponiculture has great potential in pollination and supplementary income generation.

2. Review of literature

Stingless bees originated in the earth 65 million years ago longer than honey bees and are limited to tropics and subtropics. They belong to five different genera; Melipona, Trigona, Meliponula, Dectylurina and Lestrimelitta which plays a vital role in plant pollination (Heard, 1999; Choudhari et al., 2012) with an estimation that 30% of the human food is derived from bee pollinated crops (Slaa et al., 2006). Stingless beekeeping is known as meliponiculture undertaken by traditional communities (Cortopassi-Laurino et al., 2006). Stingless bees are small to medium sized with vestigial stings, shows social level of organization and are major pollinators of flowering plants in the tropics (Ramanujam et al., 1993; Heard, 1999; Amano et al., 2000; Raju et al., 2009). Unlike most bee species that live solitary lives, the stingless bees (Apidae, Meliponini) and the true honey bees (Apidae, Apini) are highly social ('eusocial') and have complex and long lasting colonies (Michener, 1974). The stingless bee found in Kerala is *Trigona iridipennis* Smith also called 'dammer bees' locally known as 'Cherutheneecha' in Malayalam (Singh, 2013).

Trigona iridipennis Smith are kept in India for centuries for the high medicinal value of honey as well as propolis and bee wax (Rasmussen, 2013; Cortopassi-Laurino et al., 2006; Virkar et al., 2014; Kumar et al., 2012; Andualem, 2013; Choudhari et al., 2013; Choudhari et al., 2012; Rasmussen, 2013; Virkar et al., 2014). The honey fetches high price due to its less productivity as well as high demand from pharmaceutical sector (Kumar et al., 2012). Impact of anthropogenic influences on honey bees were already reported by Basavarajappa (2010). *Trigona iridipennis* Smith are seen in all places across Kerala, in natural conditions were human beings and other predators do not have easy acess. Recent studies also showed the various nesting behaviour of *Trigona iridipennis* Smith in natural habitat as well as its adaptability various antroropoegnic habitats (Virkar et al., 2014; Singh, 2013; Nair and Nair, 2001; Kumar et al., 2012; Jose and Thomas, 2012; Jose and Thomas, 2013).

A little is reported so far about the domestication of the *Trigona iridipennis* Smith as well as the aspects of Meliponiculture in Kerala. Based on these back ground, the objectives of this study were to 1) to characterize the various Meliponiculture techniques and management practices 2) the new techniques evolved for

Meliponiculture practices and management as well as value addition of stingless bee products.

3. Materials and Methods

3.1 Study area

Kerala state covers an area of 38,863 km^2 with a population density of 859 per km^2 and spread across 14 districts. The climate is characterized by tropical wet and dry with average annual rainfall amounts to 2,817 ± 406 mm and mean annual temperature is 26.8°C (averages from 1871-2005; Krishnakumar et al., 2009). Maximum rainfall occurs from June to September mainly due to South West Monsoon and temperatures are highest in May and November (Figure 1).

3.2 Study design and data collection

Hundred and twenty farmers across Kerala engaged in Meliponiculture were interviewed and their Meliponiculture practices were observed and recorded. A questionnaire was prepared and distributed among the farmers to learn their depth of knowledge in stingless bee keeping techniques and to study the current status of Meliponiculture. The plus and minus points of various stingless bee keeping techniques and management practices were identified by regular meliponairy visits. In addition, twenty five stingless bee colonies were maintained and cross checked with the techniques used by the farmers. Necessary modifications were made by trial and error method with the collective knowledge gathered from the farmers. Further improvements developed and tested in order to compensate it the collective techniques were found to be below optimum.

A meliponiary with 70 stingless bee colonies were specifically selected for the study. Three types of hives (absconded, box with inside resin coating and box without resin coating), ten numbers of each kind, were added to this meliponiary. The boxes were deployed during growth period (Sept-Dec). The boxes were kept under observation throughout the year. The nests occupied with the new swarms were noted and replaced with empty new nests of the representative type in each year.

3.3 Statistical analysis

Descriptive statistics using SPSS 12.0 (SPSS Inc., Chicago, IL, USA) were conducted to summarize the data.

4. Results and discussion

4.1 Anthropogenic habitats

Trigona iridipennis Smith is seen in forest/wild as well as in natural habitats. The major natural habitats include the living trunks of different tree species especially teak (*Tectona grandis* Linn), jack fruit (*Artocarpus integrifolius*), wild jack fruit (*Artocarpus hirsutus* Lam.), cycas (*cycas sphaerica*) and mango (*Mangifera indica* L). Usually *Trigona iridipennis* Smith occupies in tree trunks of diameter above 30 cm and levels of few centimetres above the ground. They build the nests on trees which have hollow spaces due to decay or rotting of the stem or braches which arise due to insect attack, disease as well as physical damage. In central Travancore majority of the nests in the tree trunks was seen on Maruthu (*Terminalia paniculata*) and at coastal areas it was poovarasu (*Hopea glabra*). It may be because both these trees have lot of cavities in their trunk, as farmers cut their branches regularly for mulching. In addition, the nest were also observed on telephone posts where cable box were installed, wooden master control box for power supply, air outlet pipes of latrines etc.

4.2 Domesticated habitats

The major domesticated habitats for *Trigona iridipennis* Smith across Kerala were wooden boxes (standing as well as hanging), earthen pot and vessel, garden pot, bamboo nodes, log nests, pith removed areca nut stems, earthen pipe, polyvinyl chloride (PVC) pipe, asbestos pipe, and coconut shell. The wooden boxes were made of mostly 'Maruthu' (*Terminalia paniculata*) or any hard wood of comparatively less commercial value. The size and dimensions of the wooden hives were also different. Some boxes have a dimension of 12 x 12 x 34 cm, while the hanging ones were often less dimension (10 x 10 x 36 cm). In some places farmers adopt the compartmentalization of nests, where separate chambers are provided for honey storage as well as brood rearing. Variety of materials were used for hanging the boxes which includes, ropes made out of coconut husk, plastic rope, even insulated electrical wires or steel wire. Usually the ropes were coated with neem oil, grease or used engine oil to repel ants and other walking predators to reach the nest.

The PVC pipes were also used for rearing the bees. The ends of the pipes were closed with end blocks having hole at the centre or coconut shells where the outer surface faces to outside. Holes were usually made on one of the master eye of the coconut shell which is quite easy to pierce and open to get a perfect circular lesion.

Earthen pots as well as vessels were employed for keeping *Trigona iridipennis* Smith. Usually pots of different sizes were used depending on the availability ranging from 3-5 litre capacity. The most common one have a neck size of 16 cm diameter and were usually closed using wooden plank. The ring shaped neck was tied with ropes and hanged at the eves of house or sheds.

People used to collect honey from colonies of stingless bees on trees by cutting down the portion of the timber where the bees were and later axed open. After collecting excess honey the timber was hung down from the eves of house or sheds. Later when there arouse a need for honey, this timber portion would split open to extract honey, and the splinters joined together to retain the colony. Today, however, colonies of stingless bees are reared in bamboo nodes, earthen pots, wooden boxes, coconut shells, PVC pipes etc.

4.3 Types of nests used in Meliponiculture

4.3.1 Bamboo nodes

Two or three node length of bamboo is cut out and then split open through the centre and remove the internodes in between except the internodes at both ends. Colonies of stingless bees with brood, honey and pollen are transferred into the bamboo poles. A small opening is made at one end for the bees to move in and out. Using stingless bee resin all the openings other than entrance can be sealed. Using resin a ring can made and fix around the entrance, which will help the bees in finding the door to the new nest. At sunset when the bees have all entered the nest, the bamboo pole may safely place so that it is protected from direct sun-light and rain.

4.3.2 Earthen pot

Earthen pots of different sizes were used by the farmers. At the side of the pot, an entrance is made with the help of a screw- driver for the bees to move in and out. Honey, pollen, brood and bees should be transferred into the pot and then a wooden plank may be used to cover the mouth of the pot. The crevices between the pot and the plank should be sealed up with resin. The pot is to be then hung down in a swing away from the direct sun light or rain.

4.3.3 Wooden Box

The most widely used and most convenient nest for stingless bee is wooden box. The boxes are of different dimensions, without considering the requirements of the bees and management of brood. People use any available hard wood for making the box and provide a single opening as entrance. They even nail top and bottom planks, which create difficulty during honey extraction.

4.3.4 Earthen bowl

People rear stingless bees in earthen bowls of varying size. Two bowls are arranged, one above the other. The lower is used for depositing the brood with queen, and the upper bowl for food pots, which is to be closed with wooden plank. An entrance with a resin lip is provided on the lower bowl. The upper bowl should contain an opening having 2 cm diameter for bee passage. As per the availability of honey any number of bowls with a hole at the bottom might be piled up. The new bowls thus added are placed in the middle.

4.3.5 Coconut shell

Stingless bees are also reared inside big coconut shells that are split at the middle. The bottom shells is used, the same way as in the case of earthen bowls where, the brood and queen are placed. It is closed with the upper shell after opening a single eye. Another shell with an opened eye placed downward above the bottom shell. In which food pots are deposited. All openings are sealed with resin. An entrance hole is provided at the centre of the bottom shell.

4.3.6 PVC pipe

PVC pipes of 4" diameter, split equally are in use for stingless bee rearing. Both ends of the pipe are closed using end caps or coconut shells; with an opening at one end for bee passage. PVC pipes are found to be not an ideal material for bee nest as it retains heat during summer and spoil the colony.

Figure 1. Different types of bee nests a) hives placed on the stand; b) PVC pipe nest; c) earthen pot nest; d) nest in coconut shell; e) bamboo pole nest; f) earthen bowl nest exposed; g) log nest; h) nest in garden pot; i) two tier earthen bowl nest.

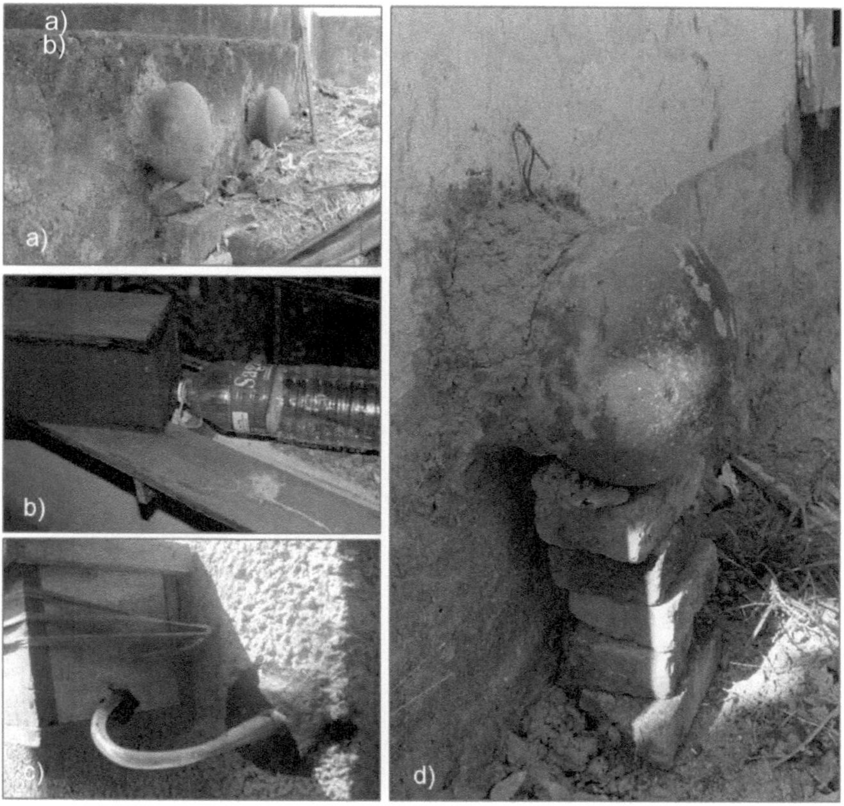

Figure 2. Colony capturing techniques a) mud pots using clay as sealant with granite support; b) collection and transfer of stingless bees using pet bottles; c) eduction using funnel and plastic tube; d) mud pots using clay as sealant with brick support.

Figure 3. Honey extraction methods a) exposing coconut shell for honey extraction; b) extraction from coconut shell; c) detached honey pots in stainless steel plate with extraction knife; d) honey pots placed on muslin cloth stretched over stainless steel utensil; e) honey pots exposed to sunlight for filtration; f) wooden box exposed for honey extraction.

Table 1. Vegetable crops as a source of nectar and pollen for*Trigona iridipennis* Smith in Kerala.

Serial number	Common name	Scientific name	Family	Source*
1	Brinjal	*Solanum melongena L.*	Solanaceae	P
2	Bitter gourd	*Momordica charantia L.*	Cucurbitaceae	P
3	Drumstick	*Morinja oleifera*	Moringaceae	NP
4	Ash gourd	*Benincasa hispida*	Cucurbitaceae	P
5	Snake gourd	*Trichosanthes cucurmerina*	Cucurbitaceae	P
6	Curry leaf	*Murraya koenigii*	Rutaceae	P
7	Chilly	*Capsicum annum*	Solanaceae	NP
8	Sponge gourd	*Luffa cylindrical*	Cucurbitaceae	P
9	Lady's finger	*Abelmoschus esculentus*	Malvaceae	P
10	Bottle gourd	*Lagenaria vulgaris*	Cucurbitaceae	P
11	Capsicum	*Capsicum frutescens*	Solanaceae	NP
12	Sweet gourd	*Momordica cochinchinensis*	Cucurbitaceae	P
13	Cheera	*Amaranthus sp*	Amaranthaceae	P
14	Pumpkin	*Cucurbito pepo*	Cucurbitaceae	P

* N= nectar, P= pollen, NP= nectar and pollen

Table 2. Fruit crops as a source of nectar and pollen for*Trigona iridipennis* Smith in Kerala.

Serial number	Common name	Scientific name	Family	Source*
1	Mango	*Mangifera indica L*	Caesalpiniaceae	NP
2	Papaya	*Carica papaya L.*	Caricaceae	P
3	Guava	*Psidium guajava L.*	Myrtaceae	P
4	Pomegranite	*Punica granatum L.*	Punicaceae	P
5	Banana	*Musa paradisica L*	Musaceae	P
6	Jack fruit tree	*Artocarpus integrifolius*	Moraceae	P
7	Rambutan	*Nephelium lappapeum*	Sapindaceae	N
8	Rose apple	*Euginea jambosa*	Myrtaceae	NP
9	Loovi	*Flacourtia inermis*	Flacourtiaceae	N
10	Carambola	*Averrhoa carambola*	Oxalidaceae	P
11	Jamun	*Eugenia cumini*	Myrtaceae	N
12	Passion fruit	*Passiflora edulis Sims.*	Passifloraceae	NP
13	Gooseberry	*Emblica officinalis*	Euphorbiaceae	P
14	Cherry	*Carissa carandas*	Apocynaceae	N
15	Bilimbi	*Averrhoa bilimbi*	Oxalidaceae	N
16	Litchi	*Litchi chinensis*	Sapindaceae	NP

* N= nectar, P= pollen, NP= nectar and pollen

Table 3. Condiments and spices as a source of nectar and pollen for*Trigona iridipennis* Smith in Kerala.

Serial number	Common name	Scientific name	Family	Source*
1	Tamarind	*Tamarindus indica L.*	Caesalpiniaceae	N
2	Cardamom	*Elettaria cardamom*	Zingiberaceae	NP
3	Cinnamon	*Cinnamomum zeylanicum*	Lauraceae	P
4	Mustard	*Brassica juncea*	Umbellifera	N
5	Coriander	*Coriandrum sativum L.*	Umbelliferae	NP

* N= nectar, P= pollen, NP= nectar and pollen

Table 4. Field crops as a source of nectar and pollen for *Trigona iridipennis* Smith in Kerala.

Serial number	Common name	Scientific name	Family	Source*
1	Tapioca	*Manihot esculenta*		NP
2	Gingelly	*Sesamum indicum L.*	Pedaliaceae	N
3	Onion	*Allium cepa*	Liliaceae	NP
4	Cotton	*Gossypium hirsutm*	Malvaceae	P
5	Jute	*Corchous olitorlus*	Tiliaceae	P
6	Pigeon pea	*Cajanus cajan*	Papilionaceae	N
7	Sunflower	*Helianthus annus L.*	Asteraceae	NP
8	Castor	*Ricinus communis L.*	Euphorbiaceae	P
9	Jetropha	*Jetropha sp.*	Euphorbiaceae	N

* N= nectar, P= pollen, NP= nectar and pollen

Table 5. Trees as a source of nectar and pollen for*Trigona iridipennis* Smith in Kerala.

Serial number	Common name	Scientific name	Family	Source*
1	Fig	*Ficus roxburghii*	Moraceae	P
2	Cannon Ball Tree	*Cauropeta guineensis*	Lecythidaceae	NP
3	Cotton tree	*Bombax malabarium*	Malvaceae	P
4	Birds cherry	*Mundingia calbura*	Verbinaceae	NP
5	Sandal	*Santalum album*	Sandalaceae	NP
6	Payyani	*Paganelia longifolia*	Bignoniaceae	P
7	Soapnut	*Sapindus emarginatus*		NP
8	Teak	*Tectona grandis Linn.*	Verbanaceae	P
9	Coper pod tree	*Peltophorum roxburghii*	Caesalpinaceae	P
10	'Ettilamaram'	*Schefflera stellata*	Araliaceae	NP
11	Bilimbi	*Averrhoa bilimbi*	Oxalidaceae	N
12	Jack tree	*Artocarpus integrifolius*	Moraceae	P
13	Oil palm	*Elacis guineensis*	Aracaceae	P
14	'Nagappoomaram'	*Couropita guianensis*	Lecithidaceae	NP
15	Bombax	*Bombax malabarium*	Malvaceae	P

* N= nectar, P= pollen, NP= nectar and pollen

Figure 4. Different types of bee nests a) absconded nest; b) new nest; c) resin coated nest.

Figure 5. Colony capturing techniques a) mud pots using clay as sealant with granite support ; b) plastic tube fitted at the entrance to prevent tunnel creation; c) eduction from wooden pillar using plastic tube and earthen pot; d) and e) eduction using funnel and plastic tube; f) collection and transfer of stingless bees using pet bottle.

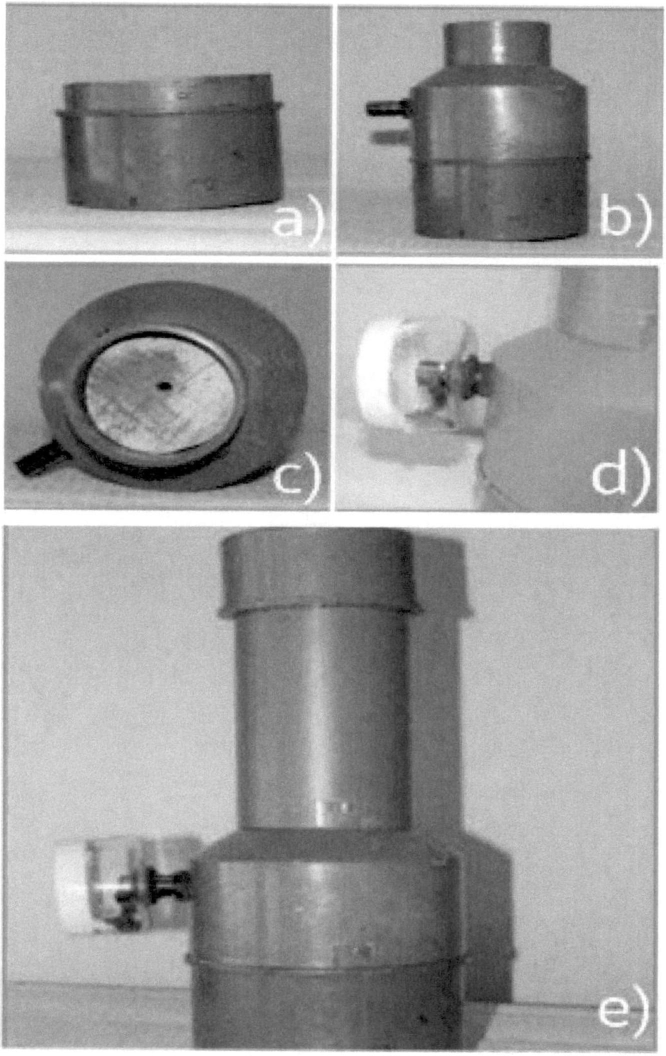

Figure 6. Detachable PVC pipe nest with separate honey chamber a) bottom brood chamber exposed; b) brood chamber with entrance tube; c) brood chamber closed with wooden piece having a hole at the centre; d) feeding bottle fitted on the entrance tube; e) PVC nest with top honey chamber.

Figure 7. Enemies of stingless bee a) crab spider capturing bees while foraging; b) red ants attacking bees; c) assassin fly attacking stingless bees; d) stingless bees trapped on spider web; e) lizard attacking stingless bees; f) waps nesting on bee hive.

Figure 8. Stingless bee worker foraging on a) *Ixora parviflora*, b) *Tagetus sp.*, c) *Artocarpus heterophyllus*, d) *Celosia sp.*, e) *Urena lobata*, f) *Euphorbia heterophylla*, g) *Anthricum andrianum*, h) *Lilium candidum*.

4.4 Methods of stingless bee colony capturing

In Kerala, most villages people are keeping stingless bees for honey. Most of the stingless bee colonies are captured from nature by the bee keepers. Stingless bee colonies are being trapped or tamed into trap boxes in different ways. Strategies for capturing and captivating bee colonies include various kinds of 'eduction' techniques (Muthuraman et al., 2013) which include trap nest and transfer of colonies by breaking the natural nest on walls and tree trunks. Mostly, stingless bee hives are positioned in stone/mud walls and tree hollows. For convenience, these nests can be divided into "can be dismantled" and "cannot be dismantled". The methods practiced by people for capturing stingless bee colonies vary, which are given below.

4.5 Hives that can be dismantled

4.5.1 Tree hollows

If the hives are in those tree hollows which could easily be dismantled, the first step is to measure the height and depth with a long twig, being inserted through the entrance of the nest. Then cut the tree trunk carefully just above and below the nest. Split the nest using an axe or chisel, transfer the queen, whole brood and pollen pots without damaging the cells and a few pots of honey, to the new hive. Close all cracks and crevices except the entrance with resin and let the bees swarm into the hive. Wait until evening to transfer the nest to a new place convenient after entering all the bees in to the nest.

4.5.2 Mud / stone wall

While capturing colonies from mud/stone walls, people take care to avoid falling of mud/stone particles which may damage the brood. The brood, queen, pollen pots and a small quantity of honey are transferred to the new hive and close all holes except the entrance. Allow the bees to enter the hive. By evening the nest is transferred to an appropriate place.

4.6 Hives that "cannot be dismantled"

Strategies for capturing hives that "cannot be dismantled" vary from the aforesaid techniques. Lack of proper accessibility is the main constraint in this type of hives. Eduction techniques are employed to overcome the issue of unaccessibility. The most commonly practiced eduction technique is the use of earthen pots which is given below.

4.6.1 Earthen pots

Stingless bee colonies can be captured from the tree hollows and walls that cannot pulled down. First, remove the entrance tube of feral colony, and then take a wide mouthed mud pot having ¾ feet height, and make a pencil sized entrance at its bottom so that the bees can easily get through it. Fix the pot in such a way that the hive's opening hole would come at the centre. The pot can be fixed safely either by nailing it or by tying it tightly with iron threads or stamping it on a stand or a stone heap. Close the mouth of the pot with clay, cow-dung or resin. It is to make sure that the bees enter and exist only through the opening hole made at the bottom of the pt. This pot is kept for six months to one year for the transfer of the colony into the pot. If the colony is transferred, shift the pot to a convenient place at dusk after covering the mouth of the pot with a wooden plank and ceiling the cracks with resin.

4.7 Trap nest preference

Various kinds of trap nests are in use for capturing stingless bee colonies. These devices are specific boxes for the purpose of trapping swarms. The various kinds of trap nests are; 1) absconded nests, 2) pre-resin coated new nest, 3) new nest (without resin coating). The preference of the stingless bee for trap nest selection is found in the following order, absconded nests> new boxes coated with resin> new box. Due to disease, infection, bad climatic conditions or due to some unknown reasons stingless bees may abandon their nest. Such absconded nests can be cleaned using wire brush of hive tool and be used as trap nests. It was found that stingless bee swarms usually prefer such absconded old nests as their new home. Newly constructed wooden boxes or nests of bamboo nodes, mud pots, coconut shells, earthen pots and earthen bowl can be used as trap nests. Cracks and cerevices of such nests are sealed and brushed with stingless bee resin. Newly constructed nests are used as trap nests to attract swarms without brushing with resins. Swarms rarely prefer these types of nests.

Figure 9. Different types of bee nests a) bee hive with partition board; b) bee veil; c) earthen bowl splitter box; d) and f) bee hive with detachable bottom and lid; e) wooden splitter box.

Figure 10. Different types of observation hives a) wooden box nest with transparent sheet; b) observation hive with glass top; c) observation hive with plastic sheet; d) exposed glass observation hive; e) glass hive with wooden cover; f) observation hive with LED lights exposed; g) LED lighted observation hive with wooden cover.

4.8 Trapping the colony using plastic bottle

At certain locations both these pot trap and plastic tube trap can't be used to collect stingless bee colonies. For example in public places like the waiting shed of a bus stop these traps can't be fixed permanently as it will be destroyed by the public. In such cases another method can be used. Take a 2 litre plastic bottle, and make small holes on it with a needle. After removing the entrance tube of the live colony, closely fit the bottle to the nest entrance after removing its lid. Then knock on the wall using a hammer gently to bring out the bees into the bottle. After collecting maximum number of bees in the bottle close the bottle and bring it to the meliponiary and take a new wooden box and put some fresh brood and food pots in the box. Take some resin and make it into a thick ring and fix it at the entrance and fit the bottle at the entrance in such a way that the bees can enter into the box and back into the bottle. All other holes should be sealed with resin. At evening all the bees will enter into the box and take care of the brood and food pots. The bottle can be removed on next day at late morning. This method will be successful if the brood contains a queen cell in it.

4.9 Method of colony division

Division of stingless bee colony is a crucial part of meliponiary management. Natural colony division/swarming can lead to reduced efficiency in meliponiary. Timely splitting (division) of colonies increase the number of stock colonies, honey yield, and prevention swarming. For propagation, production and productivity, after a specific time period, after attaining maximum growth in the nest, stingless bee swarms themselves and form new separate colonies. The new born colonies are centred by a gyne, which with half of the worker bees establish a colony in a new habitat. In stingless bee the new swarm usually do not migrate beyond an extent (Sakagami, 1982; Inoue et al., 1984; Van Veen and Sommeijer, 2000; Sommeijier et al., 2003). Colony division can be achieved by 1) conventional nests, 2) specially designed PVC nest, 3) earthen bowl, 4) coconut shell.

4.10 Use of conventional nests

Stingless bee colonies can be multiplied by splitting the colony once or twice every year. Like stinging bees, the stingless bee colony division is not so easy, as the emergency queen rearing is difficult in stingless bees. Division will be successful if only, the queen lass half contains a gyne or queen cell. To split the colony, strong colonies with enough brood and food pots are selected. Prior to splitting a new sealed hive is arranged. Half of the brood, with egg of all ages and food pots are transferred into the new nest. Take the mother colony to a distant place and place the new colony in the position of the mother colony, thus allowing the bees to enter into the new nest. Once the scattered bees enter into the new box, the division can be considered as complete.

4.11 Specially designed PVC pipe nest

Nest made up of PVC pipes and fittings are in use for division of colony. It is found to be useful for honey extraction without disturbing the queen and brood chamber. This nest consist of a bottom brood chamber and a top honey chamber. The bottom brood chamber is made up of two pieces, and an end cap at the bottom and a reducer above it, which is fitted in such a manner that they can be separated by a twisting movement. From the reducer a pipe is extended which functions as honey chamber. The honey storing pipe is covered with another PVC pipe on the top, of which a hook is screwed for hanging the nest. On the bottom chamber a tube is inserted for entrance on which a small plastic jar is fitted for artificial feeding at the time of dearth and transporting.

4.12 Earthen bowl

Earthen bowl having the size of a coconut shell is used for division of colonies. Two bowls of equal size are used. Brood, queen and food pots are deposited in the bottom bowl which contains an entrance hole at the centre, on which the second bowl is fitted in an inverted position. As the size of the nest is small, it will be filled with brood and food pots soon. At the time of splitting, detach the bowls, and if the bowls contain enough brood and food pots, attach a new set of vacant bowls to each half of the parent bowl.

4.13 Coconut shell

Stingless bee colonies reared in the coconut shell are split as described above. Large coconut shells split to equal size are used for this purpose. The brood, food pots and the queen are placed in the bottom piece which is covered with an upper piece having an entrance hole on it. When the shell is filled with brood, each piece is separated and new halves of matching size is fitted to them.

4.14 Honey extraction

In Kerala the traditional practice of honey extraction is by cutting the honey cubes using a knife or by scooping it with a steel spoon and squeezing it by hand, then filtering it through a net. The main drawback of this system is that, the fallen bees, eggs and pollen pots get squeezed and may mix with the honey, thus the honey may be get contaminated. It affects the shelf life and quality of the honey. This can be avoided by keeping the honey cubes on a net that stretched over a vessel. The honey will be normally flows down and it can be speed up by placing the whole system in slight sunlight. As the honey pots are not squeezed, the bees, eggs and pollen pots will remain as such in the net, and only pure honey will be collected in the vessel.

Honey also can be collected by placing cut honey cubes in a steel dish kept in a slanting position and keeping the honey pots at the top. The honey will be flown down and can be collected from the bottom of the dish.

4.15 Pest and disease management

Stingless bees have lot of enemies like ants, spiders, lizards, wasps, dragon fly, assassin flies, mites, solider flies, fungus, frogs, toads, termites and rats. Ants are the main enemies of stingless bees. Red ants and black ants directly attack and catch bees. Few types of ants enter into the nest and feed on the food stores, and destroy the nests. Spider construct their nests in front of the nest or on the foraging route of stingless bees, on which a lot of bees trapped. Few species of spiders show camouflage with the flowers on which the bee forages and captures them.

Lizards are harmful to stingless bee colony, as they capture bees at the nest entrance or while foraging. Wall lizards, garden lizards and skink lizard prey on stingless bees and destroy the colony. Wasps are enemies of stingless bees as they prey upon them. Dragon flies also feed on stingless bees. Assassin flies are a serious threat to the bees. Large number of stingless bee colonies is destroyed by the attack of fungus, especially nests kept in moist region. Toad and frogs are also enemies as they catch stingless bees while they are foraging. Termites attack the wooden nests and destroy the colony. Rats destroy the nests, eat pollen and brood. Usually people are not bothered about the pests or the diseases of stingless bee colonies. A good number of published works are avilabile on pest and enemies of stingless bees (Inoue et al., 1984; Roubik, 1995; Roubik, 2006; Devanesan et al., 2009).

4.16 Honey storage and use

After extracting honey, people usually store it in bottles or cans without proper processing, thus results in fermentation and spoilage. The honey is either sold as it fetches comparatively high prices than *Apis* honey, or kept for domestic use. The market price of stingless bee honey ranges from Rs 1500-2000/Kg as per locality and season. Stingless bee honey is used as home remedies for ailments like stomach disorders, respiratory ailments, burn, and even used for the treatment of cancer (Devanesan et al., 2009; Balkrishna, 2012; Sureshkumar et al., 2012).

4.17 Innovations in Meliponiculture

Since the practice of stingless bee rearing and the management of meliponiary in Kerala is not scientifically based, well organized, and properly monitored, the benefits from bee rearing is not in the optimum. A few techniques and better management practices are developed, experimented and found to be practical in the are described. Handling and colony management efficiency could be improved using various tools as well as management practices. These include eduction techniques, division methods, hygienic honey extraction, processing techniques, value addition and marketing. Tools like bee veil, ant well, honey extraction tools, and usage of observation hives could further improve the handling efficiency of the farmer. Most of these tools could easily made with less technical knowledge and expenditure.

1. Dimensions of the nesting box: it was found that people use nesting boxes of varying size without understanding the needs and biology of the colony. Continuous observations and trial and error experiments helped in standardisation. Any hard wood box with length (30 cm) x width (10 cm) x height (10 cm) (all inner measurements) is the best size for meliponicultre in Kerala. The volume of this box is ideal for brood growth, temperature regulation by the bees and hive managements. If the bottom and top board are detachable honey extraction becomes easy.
2. The specially designed PVC nests can be modified by putting an extra pipe over the honey chamber. It helps to reduce the temperature of the nest. A rain-guard can be fitted on the top of this nest, so that it can be placed outside. If a feeding bottle is fitted at the entrance, the whole unit can be transported without any trouble.

3. If a new shell is inserted in the middle in coconut shell nest, during honey flow period only honey will be stored in that chamber, which makes honey extraction easy and non destructive.
4. Ants are the most dangerous enemies of stingless bee colonies. Whenever the nest is opened there is chance for ants to enter the nest and damage the colony. This can be prevented by fitting an ant well or spreading neem oil/grease on the ropes/wires used for hanging the nest. If hives are placed on stands, an ant well at the bottom will prevent the ant entry. The main threat of stingless bee nests hanging at the eves of sheds is ants. The entry of the ants can be prevented by fixing an ant well on the swing. It can be prepared by making a small hole on a PVC end cap and passing a steel wire through it. The hole can be made water proof by pasting M-Seal (epoxy sealant from pidilite industries, Mumbai, India) in the pore. One end of the steel wire can be tied on the eve of the shed and other end to the bee hive. Water can be kept in the end cap which prevents ants and other pests. Few drops of engine oil also can be used instead of water to prevent ants and other pests.
5. During eduction using earthen pots, entry of sealing materials (mud, cow dung, clay), into the pot can be avoided by wrapping a cloth ribbon around the mouth of the pot.
6. In order to avoid the construction of an internal tunnel on the inner wall of the eduction pot, a tube of appropriate length can be fixed at the entrance of the mother nest before fitting the pot.
7. For eduction of colonies from places where a pot cannot be fitted, a plastic tube of required length attached to a wooden box can be used to overcome the difficulty.
8. Bees from feral colony which cannot be dismantled, a plastic bottle can be used to collect the bees by putting the mouth of the bottle at the entrance and tapping on the substratum. These bees can be brought into a nest which is provided with brood and food pots.
9. For easy splitting of colonies using wooden boxes a special box consisting of two pieces each having a dimension of 30 (length) x 10 (width) x 2.5 (height) cm is designed. It is designed as splitter box, which is found to be functional as the two pieces can be separated without damaging the brood and food pots.

10. For monitoring the activities of the colony during day and night, a colony observation technique is developed. Use of clear glass plates/transparent sheets is useful in observing the colonies without disturbing the activities of the bees. It also helps in easy opening of the hive as the top plank is not pasted with resin by the bees. Illuminating the hives improve the night observation of the bee colony. Red light emitting diode (LED) seems to be more comfortable as red light is not visible to bees.
11. Colony division will be a failure in the absence of a queen/gyne/queen cell. In order to make sure the emergence of a queen in the newly divided colony, queen cell grafting is found to be effective. Queen cell can be easily distinguished from other brood cells by their size. Carefully detach the queen cell from the brood cluster and place it in the new colony along with the brood. If new graft is successful a new queen will emerge and ensure the survival of the nest.
12. Bee veil: The entry of the stingless bees into hair, nostril, ear and eyes during colony handling can be prevented by using a bee veil. It can be prepared by stitching a transparent net (1 mm mesh size) on the rim of a farmer's hat. The net should hang down up to the waist. The free ends of the net should not be stitched together, but should overlap each other in such a way that it allows free movement of arms. A bee veil also can be prepared by stitching a skirt with transparent net, worn over a hat.
13. Honey processing: Fermentation of stingless bee honey due to presence of water content could be prevented by proper processing of the honey. Processing of honey can be done by heating it in a water bath, which is better than placing the honey exposed to sunlight.
14. Value addition: Value addition of honey ensure better profit to the farmers, thus enhancing their income. Value addition is feasible through the hygienic processing and management techniques. Stingless bee honey is used as an inevitable ingredient in many ayurvedic and folk medicines. Seasonal health drinks and beverages can be made using honey as a main ingredient. Honey cola can be promoted as a health drink especially during summer season in Kerala. Stingless bee honey is used as an active ingredient in many of the cosmetic preparations (Devanesan et al., 2009; Sureshkumar, 2012). Various fruits can be preserved in honey for long periods.

15. Proper packing and marketing through established outlets are also important. This can be done by farmer's co-operatives.

The Stingless bee nests are great example of colonial life of social insects, which plays a major role protecting from various environmental factors as well as from predators (Pavithra et al., 2012). In addition it provides a microclimate, storage of honey, pollen and rearing young ones. The selection of various natural and adapted anthropogenic habitats may suggest their increasing adaptability and easy rearing in meliponiculture. There is not much recorded evidence about the cultivation history of *Trigona iridipennis* Smith in Kerala, however it may concluded that their culture is popular in rural as well as some tribal areas. The main reason may be the easiness in keeping as well as their high demand in ayurvedic, unani and traditional medicinal preparations.

Workers make wax that is secreted from dorsal glands (Michener, 1974; Sakagami, 1982). Different stingless bee waxes have different chemical properties, and are much simpler than wax of Apis (Blomquist et al., 1985). Cerumen, a product of plant resin mixed with wax and employed exclusively by meliponines, remains soft for an extended time and are more pliable than bees wax that of *Apis mellifera* (Hepburn and Kurstjens, 1984).

Various beekeeping methods preferred by farmers across Kerala for the cultivation of *Trigona iridipennis* Smith. Each nest has its own advantage and disadvantage. During the survey, the most preferred one's were wooden box. Even then according to the easy availability and production cost different nests like earthen pot, bamboo nodes, coconut shell, PVC pipes etc were used. The most preferred natural nesting sites by *Trigona iridipennis* Smith were mud/stone walls in our study. The closeness to the food availability as well as the rigid nature of the habitat may be a reason for the selection of these habitats. The adaptation to various habitats can be governed by environmental factors in addition to the genetic factors (Lima and Dill, 1990; Jaenike and Holt, 1991; Martin, 2001; Cameron et al., 2004; David, 2006). Therefore a better understanding of the genetic nature of *Trigona iridipennis* Smith may able to explain the attractiveness of various habitats.

In earthen bowl and coconut shell methods it is very advantageous to collect honey without causing any disturbance to the brood chamber that is at the bottom. In PVC pipe and coconut shell the colony often desert in summer due to high temperature.

Even though the bamboo nodes are alright, they are found to degenerate after few years through the attack of wood borers.

5. Conclusions

Stingless bee rearing is an important activity as it helps pollination and provides honey for various purposes. The bottle neck for Meliponiculture includes availability of colonies to start with, better techniques for colony capture, colony multiplication, honey extraction, processing and marketing. This study highlights the various drawbacks and flaws of traditional Meliponiculture in Kerala, which could be rectified using improved scientific management practices and tools. If correctly informed disseminated these techniques could be utilized by tribal, woman, old aged people and even children to fetch an extra income and also to serve the high demand of stingless bee honey and help to maintain the rich biodiversity of Kerala. Being a social insect, *Trigona iridipennis* Smith shows great diversity in nesting pattern in natural habitats as well as anthropogenic habitats. The shift towards anthropogenic habitats than natural ones may be due to destruction of natural habitats as well as the availability of manmade habitats. Our study also highlights the various cultivation feasibility of *Trigona iridipennis* Smith culture which can be explored more for the demand of stingless bee honey (Dammer honey) in Kerala.

References

Amano, K., Nemoto, T. & Heard, T.A. (2000). What are stingless bees, and why and how to use them as crop pollinators?-a review. *Japan Agricultural Research Quarterly*, *34*(3), 183-190.

Andualem, B. (2013). Synergistic antimicrobial effect of Tenegn honey (*Trigona iridipennis*) and garlic against standard and clinical pathogenic bacterial isolates. *International Journal of Microbiological Research*, *4*(1), 16-22.

Armor, M. (2005). Stingless bees in Angola. *Stingless-bees-in-angola. (accessed on 25 February/2015).*

Ayala, R., Gonzalez, V. H., & Engel, M. S. (2013). Mexican stingless bees (Hymenoptera: Apidae): Diversity, distribution, and indigenous knowledge. In *Pot-Honey* (pp. 135-152). Springer New York.

Balkrishna, A. (2014). A practical approach to the science of Ayurveda-A comprehensive guide for healthy living, Divya Yog Mandir Trust, Haridwar, India.

Bänziger, H., Pumikong, S. & Srimuang, K.O. (2011). The remarkable nest entrance of tear drinking *Pariotrigona klossi* and other stingless bees nesting in limestone cavities (Hymenoptera: Apidae). *Journal of the Kansas Entomological Society*, *84*(1), 22-35.

Basavarajappa, S. (2010). Studies on the impact of anthropogenic interference on wild honeybees in Mysore District, Karnataka, India. *African Journal of Agricultural Research*, *5*(4), 298-305.

Biesmeijer J.C., Guirfa M., Koedam D., Potts S.G., Joel D.M., & Dafni A. (2005). Convergent evolution: floral guides, stingless bee nest entrances, and insectivorous pitchers, *Naturwissenschaften, 92*(1), 444-450.

Boorn, K.L., Khor, Y.Y., Sweetman, E., Tan, F., Heard, T.A., & Hammer, K.A. (2010). Antimicrobial activity of honey from the stingless bee Trigona carbonaria determined by agar diffusion, agar dilution, broth microdilution and time-kill methodology. *Journal of Applied Microbiology*, *108*(5), 1534-1543.

Cameron, E.C., Franck, P. & Oldroyd, B.P. (2004). Genetic structure of nest aggregations and drone congregations of the southeast Asian stingless bee *Trigona collina*. *Molecular Ecology*, *13*(8), 2357-2364.

Camrgo J.M.F., & Pedro S.R.M. (1992). Systematics, phylogeny and biogeography of the Meliponinae (Hymenoptera, Apidae): a mini- review. *Apidologie, 23*(1), *509-522.*

Chinh, T.X., Sommeijer, M.J., Boot, W.J., & Michener, C.D. (2005). Nest and colony characteristics of three stingless bee species in Vietnam with the first description of the nest of *Lisotrigona carpenteri* (Hymenoptera: Apidae: Meliponini). *Journal of the Kansas Entomological Society, 78*(4), 363-372.

Choudhari, M.K., Haghniaz, R., Rajwade, J.M. & Paknikar, K.M. (2013). Anticancer activity of Indian stingless bee propolis: an in vitro study. *Evidence-Based Complementary and Alternative Medicine, 2013*(2013), 1-10.

Choudhari, M.K., Punekar, S.A., Ranade, R.V., & Paknikar, K.M. (2012). Antimicrobial activity of stingless bee (*Trigona sp.*) propolis used in the folk medicine of Western Maharashtra, India. *Journal of Ethnopharmacology, 141*(1), 363-367.

Choudhari, M.K., Punekar, S.A., Ranade, R.V., & Paknikar, K.M. (2012). Antimicrobial activity of stingless bee (*Trigona* sp.) propolis used in the folk medicine of Western Maharashtra, India. *Journal of Ethnopharmacology, 141*(1), 363-367.

Cortopassi-Laurino, M., Imperatriz-Fonseca, V. L., Roubik, D. W., Dollin, A., Heard, T., Aguilar, I., & Nogueira-Neto, P. (2006). Global meliponiculture: challenges and opportunities. *Apidologie, 37*(2), 275.

Cortopassi-Laurino, M., Imperatriz-Fonseca, V.L., Roubik, D.W., Dollin, A., Heard, T., Aguilar, I., Venturiei, G.C., Eardley, C. & Nogueira-Neto, P. (2006). Global meliponiculture: challenges and opportunities. *Apidologie, 37*(2), 275-292.

Danaraddi, C.S., & Viraktamath, S. (2007). Studies on stingless bee, *Trigona iridipennis* Smith with special reference to foraging behaviour and melissopalynology at Dharwad, Karnataka. *Master of Science Thesis. College of Agricultural Science. Dharwad.*

Danaraddi, C.S., Viraktamath, S., Basavanagoud, K. & Bhat, A.R.S. (2010). Nesting habits and nest structure of stingless bee, *Trigona iridipennis* Smith at Dharwad, Karnataka. *Karnataka Journal of Agricultural Sciences, 22*(2), 310-313.

David, W.R. (2006). Stingless bee nesting biology. *Apidologie, 37*(2), 124-143.

DeMera, J.H., & Angert, E.R. (2004). Comparison of the antimicrobial activity of honey produced by *Tetragonisca angustula* (Meliponinae) and *Apis mellifera* from different phytogeographic regions of Costa Rica. *Apidologie*, *35*(4), 411-417.

Devanesan S., Shailaja K. K., & Pramila K. S. (2009). Status paper on stingless bee *Trigona iridipennis* Smith. All India Co-ordinated Research Project on Honey bees and pollinators, Vellayani Centre,Thiruvananthapuram.pp79.

Drumond, M.P., Zucchi R., Yamane, S., & Sakagami S.F. (1998). Oviposition behaviour of the stingless bees XX. *Plebeia (Plebeia) juliani* which forms very small brood batches (Hymenoptera:Apidae,Meliponinae). *Entomological Science*, *1*(2), 195-205.

Franck, P., Cameron, E., Good, G., Rasplus, J.Y., & Oldroyd, B.P. (2004). Nest architecture and genetic differentiation in a species complex of Australian stingless bees. *Molecular Ecology*, *13*(8), 2317-2331.

Heard, T.A. (1999). The role of stingless bees in crop pollination. *Annual Review of Entomology*, *44*(1), 183-206.

Inoue, T., Sakagami, S. F., Salmah, S., & Yamane, S. (1984). The process of colony multiplication in the Sumatran stingless bee *Trigona* (Tetragonula) *laeviceps*. *Biotropica*, 100-111.

Jaenike, J. & Holt, R.D. (1991). Genetic variation for habitat preference: evidence and explanations. *American Naturalist, 137*(1), S67-S90.

Jose, S.K., & Thomas, S. (2012). Stingless beekeeping (Meliponiculture) in Kerala. In: Selected beneficial and harmful insects of Indian subcontinent, (Thomas, K.S (2012) ed. LAP LAMBERT Academic Publishing GmbH & Co. KG, Saarbruken, Germany.

Jose, S.K., & Thomas, S. (2013). Nest architecture of *Trigona iridipennis*. Proceedings of the National Seminar on Invertebrate Taxonomy. Nirmala Academic and Research Publications, Kerala, India.

Kent, R.B. (1984). Mesoamerican stingless beekeeping. *Journal of Cultural Geography*, *4*(2), 14-28.

Kolmes, S.A., & Sommeijer, M.J. (1992). Ergonomics in stingless bees: changes in intranidal behaviour after partial removal of storage pots and honey in *Melipona favosa* (Hym. Apidae, Meliponinae). *Insects Sociaux, 39*(2), 215-232.

Krishnakumar, K.N., Prasada Rao, G.S.L.H.V., & Gopakumar, C.S. (2009). Rainfall trends in twentieth century over Kerala, India. *Atmospheric Environment, 43*(11), 1940-1944.

Kumar, M.S., Singh, A.J.A.R., & Alagumuthu, G. (2012). Traditional beekeeping of stingless bee (*Trigona* sp) by Kani tribes of Western Ghats, Tamil Nadu, India. *Indian Journal of Traditional Knowledge, 11*(2), 342-345.

Lima, S.L., & Dill, L.M. (1990). Behavioural decisions made under the risk of predation: a review and prospectus. Canadian Journal of Zoology, *68*(4), 619-640.

Marisa, H., & Salni, S. (2012). Red wood (*Pterocarpus indicus* wild) and bread fruit (*Artocarpus communis*) bark sap as attractant of stingless bee (*Trigona* spp). *Malaysian Journal of Fundamental and Applied Sciences, 8*(2), 107-110.

Martin, T.E. (2001). Abiotic vs. biotic influences on habitat selection of coexisting species: climate change impacts?. *Ecology, 82*(1), 175-188.

Michener, C.D. (1974). The social behaviour of the bees. A comparative study. Belknap, Harvard University Press, Cambridge, Mass.404pp.

Mohan, R. & Devanesan, S. (1999). Dammer bees, *Trigona iridipennis* Smith (Apidae: Meliponinae) in Kerala. *Insect Environment, 5*, 79.

Muthuraman, M., Jose, S.K., Mohankumar, S., & Vijayakumar, K. (2013). Bioecology of stingless bee, *Trigona iridipennis Smith* and evolving improved Meliponiculture techniques. Monograph on Morphometry and Phylogepgraphy of honey bees and stingless bees in India, University of Agricultural Sciences, Dharward, India.

Nair, M.C. (2003). Apiculture resource biodiversity and management in Southern Kerala. PhD. Thesis. Mahatma Ghandhi University, Kottayam, 277.

Nair, M.C., & Nair, P.K.K. (2001). Beekeeping by Kanikkars in southern Western Ghats of Kerala. *Indian Bee Journal, 63*(1 & 2), 11-16.

Pavithra, N., Shankar, R., & Jayaprakash. (2012). Nesting pattern preferences of stingless bee, *Trigona iridipennis* Smith (Hymenoptera: Apidae) in Jnanabharathi campus, Karnataka, India. *International Research Journal of Biological Sciences, 2*(2), 44-50.

Raju, A.J.S., Rao, K.S., & Rao, N.G. (2009). Association of Indian stingless bee, *Trigona iridipennis* Smith (Apidae: Meliponinae) with Red-listed *Cycas sphaerica* Roxb. (Cycadaceae). *Current Science, 96*(11), 1435-1436.

Ramanujam, C.G.K., Fatima, K., & Kalpana, T.P. (1993). Nectar and pollen sources for dammer bee (*Trigona iridipennis* Smith) in Hyderabad (India). *Indian Bee Journal*, *55*(1/2), 25-28.

Rasmussen, C. (2013). Stingless bees (Hymenoptera: Apidae: Meliponini) of the Indian subcontinent: Diversity, taxonomy and current status of knowledge. *Zootaxa*, *3647*(3), 401-428.

Roubik, D. W. (1995). *Pollination of cultivated plants in the tropics* (No. 118). Food & Agriculture Organization, USA.

Roubik, D.W. (1989). Ecology and natural history of tropical bees. Cambridge university Press, New York.514p.

Roubik, D.W. (2006). Stingless bee nesting biology. *Apidologie, 37*(2).124-143.

Sakagami, S. F. (1982). Stingless bees. *Social insects*, *3*, 361-423.

Singh, R.P. (2013). Domestication of *Trigona iridipennis* Smith in a newly designed hive. *National Academy Science Letters*, *36*(4), 367-371.

Sommeijer, M.J., Beuvens, F.T. & Verbeek, H.J. (1982). Distribution of labour among workers of *Melipona favosa* F: Construction and provision of brood cells. *Insectes Sociaux,* 29(2), 222-237.

Sommeijer, M.J., De Bruijn, L.L.M., Meeuwsen, F.J.A.J., & Martens, E.P. (2003). Natural patterns of caste and sex allocation in the stingless bees *Melipona favosa* and *M. trinitatis* related to worker behaviour. *Insectes Sociaux*, *50*(1), 38-44.

Suresh Kumar, M., Singh, R.A., & Alagumuthu, G. (2012). Traditional bee keeping of stingless bee (*Trigona sp.*) by Kani tribes of Western Ghats, Tamil Nadu, India. *Indian Journal of Traditional Knowledge*, *11*(2), 342-345.

Van Veen, J.W., & Sommeijer, M.J. (2000). Colony reproduction in *Tetragonisca angustula* (Apidae, Meliponini). *Insectes sociaux*, *47*(1), 70-75.

Virkar, P.S., Shrotriya, S., & Uniyal, V.P. (2014). Building walkways: observation on nest duplication of stingless bee *Trigona iridipennins Smith. Ambient Science,* 1(1), 38-40.